Language for maths

y3

smaller

Altogether

Greater

combined

Share

each

less

increase

By Steve Cooke & Graham Smith, The EAL Academy

Contents

How to use this book

This book is divided into five sections: Number – number and place value; Number – addition and subtraction; Number – multiplication and division; Number – fractions; Explain why.

Each section consists of: instructions for a number of collaborative activities; a range of sample questions (based on papers 2 and 3 from the 2016 and 2017 SATs) plus worked support examples to guide pupils, and answers for all questions. Accompanying printable resources and answers for the collaborative activities are available via the free online download.

Pupils should undertake the collaborative activities first in order to rehearse the language needed when answering the supported questions.

Collaborative activities

Collaborative activities are learning activities which are designed for the purpose of getting pupils to understand and use vocabulary and sentence structures which are relevant and important to the subject. They are also intended to enable pupils to better understand the concepts and processes involved in solving problems. In this way, the language and content of the subject matter are integrated.

There are a number of different strategies for creating opportunities for pupils to understand and use language in the context of mathematics. Each of the strategies creates the need for pupils to ask each other questions and make statements about the topic.

Card chains, card sequences and domino chains

These activities are intended to enable pupils to link previous information to new information and to get pupils to calculate on the basis of the new information. Card chains can also be used to encourage pupils to think about sequence. For example, they encourage pupils to think and talk about the order of numbers in terms of lowest to highest.

Card sorting

These activities require pupils to sort cards into categories. For example, they may require pupils to sort the cards into the appropriate columns on a grid or the appropriate section on a sorting table. They encourage pupils to differentiate between cards which contain similar but different information.

Card matching

These activities support pupils in arranging a set of cards on a grid so that the information on the cards matches across a row. For example, they may involve matching a question card with an answer card or a calculation card with an explanation card.

Clue sheets

Clue sheets require pupils to read and share clues which help them to complete a table or diagram. The clue sheets have different clues on them so that pupils have to cross-reference clues in order to complete the task. A task for four pupils may involve four different clue sheets so that each pupil has to contribute.

Information gap / barrier games

Information gap games involve pupils being given some information, but they will also be missing some important information. So Pupil A may have half the information needed to complete a task while Pupil B has the other half. This creates the need for pupils to ask and answer each other's questions in order to complete the task. For example, Pupil A may have half of the bars on a bar chart and will have to ask Pupil B questions so that he / she can fill in the missing bars on the chart accurately and, at the same time, Pupil B will have to ask Pupil A questions to obtain the missing information that he / she needs.

Four in a line games

These activities are useful for practising vocabulary and sentence structures and encouraging flexibility of thinking. There are a number of different rectangles on a game board where each supplied card could be placed. Rather than simply choosing the first possible rectangle, pupils need to scan the board to see if they can find a rectangle which gives a greater advantage in terms of connecting four of their cards and thereby winning the game. The game-like nature of the activity encourages natural repetition of vocabulary and sentence structures and also often means that pupils want to 'play again'.

Substitution tables

Substitution tables are useful for supporting pupils' abilities to form relevant questions and statements. Pupils generate a range of questions for a partner to answer. Pupils could also work together to form questions for another group or pair. This requires them to be sure that they know the correct answers to their questions. Pupils can also devise statements and give them to other pupils to check. A variation on this is to ask pupils to devise a set of statements, some of which are true and some of which are false. Another group of pupils can then check the statements and change the false statements to make them true. Substitution tables can be used to encompass a wide range of topics to familiarise pupils with typical vocabulary and sentence structures relating to these topics.

Collaborative guided talk

These activities aim to engage pupils in a small-group dialogue which is initially led by a teacher or teaching assistant. The adult asks questions and provides prompts to shape the thinking and language around a particular problem. This also involves modelling the appropriate language and re-casting what pupils say. As the activity proceeds, the intention is for pupils to 'take over' the questioning and engage in a greater proportion of the dialogue.

Number – number and place value

Collaborative activities

Information gap / barrier game: Counting in multiples of 4, 8, 50 and 100

An activity for two pupils. *See download for accompanying printable resources.*

Content: Counting in multiples of 4, 8, 50 and 100

Key language

The **second** number in the **first row** is 4. The **fifth** number in the **second row** is 56.

If you **count on in fours / eights / fifties, the next number** will be 12 / 24 / 150.

If you are **counting in multiples of 4** and you are on 16 then **the next number** will be 20.

There is a number **missing**. The **missing number** is 20. 20 **should be between** 16 **and** 24.

Instructions

1. Pupil A has sheet A and Pupil B has sheet B.

2. Pupil A has information on their sheet which Pupil B does not have and Pupil B has information which Pupil A does not have.

3. Pupils take turns to say aloud the numbers they have in the first row on their sheet (e.g. "The third number in the first row is 8.") and to write the missing numbers on their sheets.

4. When they have written all the numbers in the first row, they decide together if a number is missing from the sequence. They then decide together what the missing number is and where it should be in the sequence (i.e. between 16 and 24) and write this on their sheets.

5. Pupils carry out the same procedure to complete the rest of the rows on the sheets.

6. When they have finished, the pupils complete the writing sequence sheet by writing the numbers in the correct sequence. They should each have a writing sequence sheet.

Card sorting: Compare and order numbers up to 1000

An activity for three pupils. *See download for accompanying printable resources.*

Content: Comparing and ordering numbers up to 1000

Key language

437 is **more than** 347 **but less than** 473.

743 is **the largest** number.

347 is **the smallest number**.

Instructions

1. Cut out the digit cards. Each pupil has two sets of the digit cards so that each group of three pupils has six sets of cards altogether and a copy of the board.

2. Pupil A chooses three of their digit cards from one of their sets and uses them to make a three-digit number (e.g. 256), which they place on the desk.

3 Pupil B uses the same digits from one of their sets to make a different number (e.g. 562) and places it on the desk.

4 Pupil C uses the same digits from one of their sets to make a different number (e.g. 625) and places it on the desk.

5 Each pupil then uses the same digits from their second set of digit cards to make another different number and places it on the desk.

6 The pupils then decide together the order of their numbers from largest to smallest and arrange them on the board to show this.

7 They then write the numbers in the correct order on the recording sheet.

8 Repeat the activity with three different digit cards.

Four in a line: 10 or 100 more than or less than

An activity for two pupils, or four pupils in two teams of two. *See download for accompanying printable resources.*

Content: Finding 10 or 100 more than or less than a given number

Key language

446 is **ten more than** 436. 546 is **one hundred more than** 446.

436 is **ten less than** 446. 236 is **one hundred less than** 336.

What number is **ten / one hundred more than / less than** ...?

Instructions

Note: The board and sheets of cards should be enlarged to A3 size.

1 Cut the cards into individual cards.

2 One pupil or team has the yellow cards and the other pupil or team has the green cards.

3 Place the cards face down in two piles.

4 The yellow team takes the top card from their pile and places it on a correct box on the board. When they put the card down, they must say a correct sentence (e.g. "446 is ten more than 436.").

5 The green team then places their card and says a sentence.

6 The aim of the game is to get four cards of your colour in a line – across, down or diagonally.

7 The winner is the first pupil or team to get four of their colour in a line.

Clue sheets: Roman numerals

An activity for two pupils. *See download for accompanying printable resources.*

Content: Using Roman numerals from I to XII

Key language

11 is **XI** in Roman numerals. **VI** is **6** in figures.

Kamil went to a piano lesson at **10 a.m. 10** is **X** in Roman numerals.

Context

Ten different children have ten different activities starting at different times. Using the clues, the pupils have to place the cards around the clock (which has Roman numerals) to show the correct time of each child's activity.

1. Pupil A has clue sheet A and Pupil B has clue sheet B. The pupils have a copy of the board each, and a spare board and a set of cards between them.

2. Pupils take turns to read a clue from their sheet and decide together where the appropriate card should be placed on the spare board so that it shows the correct time of the child's activity.

3. When they have correctly placed all the cards, the pupils use their own copy of the board to write the activity and the time in the boxes.

Card matching: Reading and estimating the position of numbers on a number line up to 1000

An activity for two, three or four pupils. *See download for accompanying printable resources.*

Content: Reading numbers up to 1000 in numerals and in words and estimating the position of numbers on a number line

Key language

312 is three hundred and twelve.

Four hundred and twenty-four is between four hundred and five hundred.

One hundred and seventy-seven is nearer to two hundred than it is to one hundred.

Seven hundred and fifty-two is about halfway between seven hundred and eight hundred.

Note: There are three different activities:

1. Matching numbers in words to numbers in numerals

2. Matching cards with numbers in numerals to places on the number line

3. Matching cards with numbers in words to places on the number line.

Instructions

1. Cut out the cards.

2. Share out the numbers in words cards among the pupils and place the numerals cards face up on the table.

3. Pupils take turns to place the numerals cards in the correct order on the number line matching board.

4. Pupils then take turns to read aloud one of the numbers in words cards and match the card to the correct numerals card on the number line matching board.

5. When they have matched all the cards, each pupil completes a recording sheet, writing the numbers in order from smallest to largest.

6. Pupils repeat the activity, this time placing the numbers in words cards on the board.

Substitution tables: Number and place value

See download for full-size printable tables.

Content: Asking questions and making statements about place value

Key language

See substitution tables.

Asking questions

Use the table to create five questions. Ask your friend to answer your questions. Check to see if their answers are correct.

How many	ones / tens / hundreds	are there in	

nine hundred and sixty-five?

six hundred and forty-two?

four hundred and thirty?

one hundred and ninety-three?

seven hundred and thirty-nine?

four hundred and sixteen?

three hundred and seventy-eight?

two hundred and fifty-seven?

five hundred and twenty-one?

seven hundred and two?

Making statements

Use the table to create five correct sentences. Ask your friend to check them to make sure they are true.

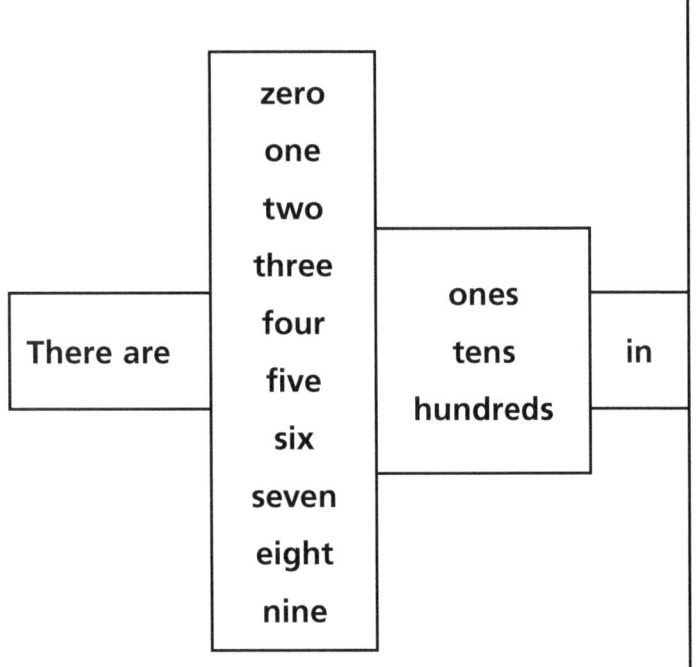

There are	zero / one / two / three / four / five / six / seven / eight / nine	ones / tens / hundreds	in	

nine hundred and sixty-five.

six hundred and forty-two.

four hundred and thirty.

one hundred and ninety-three.

seven hundred and thirty-nine.

four hundred and sixteen.

three hundred and seventy-eight.

two hundred and fifty-seven.

five hundred and twenty-one.

seven hundred and two.

Number – number and place value

Questions with support

1 Write these Roman numerals in figures. One has been done for you.

a.

| III | 3 |

b.

| IV | |

c.

| VIII | |

d.

| XI | |

 **Write the Roman numerals in figures** means write them in **numbers**, for example, 9.

Remember that **X** is 10, **V** is 5 and **I** is 1.

You also need to know that if **a letter for a smaller number** is **before a letter for a larger number** you have to **subtract** the smaller number from the larger number. Therefore:

$$\text{IX is } 10 - 1 = 9$$

2 Anna has three activities today. The clocks show the time of the three activities.

What time does each clock show? Write the time in figures in the boxes.

Dancing	Football	Music
a.m.	p.m.	p.m.

 **Write the time in figures** means write the time in **numbers**, for example, 11 a.m.

3 **a.** What number is 100 more than 229?

b. What number is 100 less than 627?

c. What number is 10 more than 884?

d. What number is 10 less than 550?

> Remember that **more than** in this kind of question means **add**. Therefore, 100 **more than** 273 is 273 **plus** 100. 100 **more than** 273 is 373. 10 **more than** 216 is 216 **plus** 10. Therefore, 10 **more than** 216 is 226.
>
> **Less than** in this kind of question means **subtract**. Therefore, 100 **less than** 752 is 752 **minus** 100. Therefore, 100 **less than** 752 is 652.

4 Fill in the missing numbers in the table.

100 less than	Number	100 more than
248	348	
	482	
		617

> In another example, 100 **less than** 387 is 387 **minus** 100 (287).
>
> 100 **more than** 387 is 387 **plus** 100 (487).

5 Write these numbers in order. Start with the smallest number.

706   559   408   599   294

Smallest   **Largest**

> When ordering numbers, it can be useful to write the numbers in columns. It may help to write the numbers in a table like this, which shows the **place value** of each **digit**.
>
> | Place value | | |
> |---|---|---|
> | Hundreds | Tens | Ones |
> | | | |
>
> Look at the **hundreds** first. Is there a number with **fewer hundreds** than the other numbers? This is the **smallest** number. Look at the **hundreds** again. Is there a **second smallest** number of **hundreds**?
>
> Are there two numbers with the **same** number of **hundreds**? If there are, then look at the **tens**. Does one of the numbers have **fewer tens** than the other one? This is the **next smallest** number.

6 A shop has these four computers for sale.

Name of computer	Price
Super	£457
Aki	£650
Jet	£590
Win	£515

a. Which computer has the highest price?

b. Which computer has the lowest price?

 The **highest** price is the **largest** number. The **lowest** price is the **smallest** number.

Write the names of the computers, **not the prices**, in the boxes.

7 Look at these six numbers.

718   472   436   264   357   243

a. Which number has three hundreds?

b. Which number has seven hundreds?

c. Which number has four tens?

d. Which number has six ones?

 Use your place value table again to help you.

Place value		
Hundreds	Tens	Ones

8 Laura is talking about one of these numbers. Which number is she talking about? Write the correct number in the box.

473

344

374

347

It has seven tens.

It has four ones.

It has three hundreds.

9 Write these numbers in numerals.

a. Six hundred and thirty-seven

b. Nine hundred and forty-five

c. Three hundred and eight

Use this table to help you.

1	2	3	4	5	6	7	8	9
one	two	three	four	five	six	seven	eight	nine

10	11	12	13	14	15	16	17	18	19
ten	eleven	twelve	thirteen	fourteen	fifteen	sixteen	seventeen	eighteen	nineteen

20	30	40	50	60	70	80	90	100	200
twenty	thirty	forty	fifty	sixty	seventy	eighty	ninety	one hundred	two hundred

For example:

Two hundred and forty-eight is:

two hundred: 200

and forty: 40

eight: 8

Therefore, two hundred and forty-eight is 248 in numerals.

10 Write these numbers in words.

a.

472

b.

129

c.

853

> ❗ Remember that when you write a number in words you must write **and** between the hundreds words and the tens words.
>
> So 536 in words is five hundred **and** thirty-six.

11 **a.** Paulina is counting in multiples of 50. Fill in the missing numbers.

50	100			250		

> ❗ In another example, Paulina is counting **in multiples of 10**. The numbers in this sequence **increase by 10** each time.
>
> | 10 | 20 | 30 | 40 | 50 | 60 | 70 |
> |---|---|---|---|---|---|---|
>
> The numbers are **10 more** than the number before.
>
> Therefore, 10 + 10 = 20, 20 + 10 = 30, 30 + 10 = 40 and so on.

b. The numbers in this sequence increase by the same amount each time. Fill in the missing numbers.

0	8		24			

12 Which of these numbers is **not** a multiple of 4? Circle it.

16		18		24		36

> ❗ **Count on** from zero in **multiples of 4**, like this, 0, 4, 8, …
>
> Which of the numbers in the boxes do you **not** count?

Answers

1. **a.** 3   **b.** 4   **c.** 8   **d.** 11

2. Dancing: 10 a.m.   Football: 4 p.m.   Music: 6 p.m.

3. **a.** 329   **b.** 527   **c.** 894   **d.** 540

4.

100 less than	Number	100 more than
248	348	**448**
382	482	**582**
417	**517**	617

5.

294	408	559	599	706

Smallest   **Largest**

6. **a.** Aki   **b.** Super

7. **a.** 357   **b.** 718   **c.** 243   **d.** 436

8. 374

9. **a.** 637   **b.** 945   **c.** 308

10. **a.** Four hundred and seventy-two

    b. One hundred and twenty-nine

    c. Eight hundred and fifty-three

11. **a.**

50	100	**150**	**200**	250	**300**	**350**

    b.

0	8	**16**	24	**32**	**40**	**48**

12. 18

Number – addition and subtraction

Collaborative activities

Card matching: Addition

An activity for two pupils. *See download for accompanying printable resources.*

Content: Mentally adding three-digit numbers to tens and hundreds

Key language

Are there two numbers which **add up to** 376? 346 **plus** 30 equals 376.

Which two numbers **add up to** 376? 346 and 30 **add up** to 376.

376 is 30 **more than** 346.

Instructions

There are two similar activities. Board 1 and the cards for board 1 involve adding three-digit numbers to tens. Board 2 and the cards for board 2 involve adding three-digit numbers to hundreds. Pupils should complete board 1 first.

1. Give the pupils a copy of board 1 and a set of cards for board 1.
2. Share out the cards between the pupils.
3. The pupils work together to make number sentences that give the correct totals (e.g. 346 plus 30 equals 376).
4. Prompt the pupils to explain why the number sentence is correct using the key language above.
5. When they have completed board 1 correctly, the pupils should repeat the activity using board 2 and the board 2 cards.

Information gap / barrier game: Column addition and subtraction

An activity for two pupils. *See download for accompanying printable resources.*

Content: Interpreting place value strips, using column addition to add and subtract three-digit numbers, and using the inverse operation to check calculations

Key language

The number has three **hundreds**, four **tens** and six **ones**. The number is **three hundred and forty-six**.

346 **plus** 242 **equals** 588. 242 **plus** 346 also **equals** 588.

588 **minus** 346 **equals** 242. 588 **minus** 242 **equals** 346.

If 346 **plus** 242 **equals** 588, **then** 588 **minus** 242 must **equal** 346.

Instructions

The term 'inverse' can be introduced to the pupils as appropriate.

1. Cut out the place value strips.
2. Give Pupil A numbers 1, 3 and 5 of the place value strips and give Pupil B numbers 2, 4 and 6.
3. Give each pupil a copy of the column addition and subtraction templates.

4 Pupil A takes place value strip 1 and says aloud the hundreds, tens and ones of the first number: "The number has three hundreds, four tens and six ones." Both pupils write the number in the top row of the first column addition template for question 1. Pupil A then says aloud the hundreds, tens and ones of the second number on the place value strip: "The number has two hundreds, four tens and two ones." Both pupils write the number on their template.

	hundreds	tens	ones
	3	4	6
+	2	4	2

5 Both pupils add the two numbers and check each other's answers. They have to agree the correct total.

6 Both pupils then use the second column addition template for question 1 and write the second number from the place value strip in the first row and the first number in the second row. They add the two numbers and again check each other's totals.

	hundreds	tens	ones
	3	4	6
+	2	4	2
	5	8	8

	hundreds	tens	ones
	2	4	2
+	3	4	6
	5	8	8

7 Pupils then write the answer to the addition calculation in the first row of both column subtraction templates. They then write the first number from the place value strip in the second row of the first subtraction template and the second number in the second row of the other subtraction template.

	hundreds	tens	ones
	5	8	8
−	3	4	6
	2	4	2

	hundreds	tens	ones
	5	8	8
−	2	4	2
	3	4	6

8 Pupils complete these inverse calculations to check their addition answers.

9 Pupil B then takes place value strip 2 and they follow the same procedure. They repeat for all six strips.

Clue sheets: Word problems

An activity for two or four pupils. *See download for accompanying printable resources.*

Content: Solving addition and subtraction word problems

Key language

How much did the computer game **cost**? The computer game **cost** £21.

How much money **did** she **have**? She **had** £45.

How much money **does she have now**? **Now she has** £24.

Context

Eight different people went to the shops and bought eight different items. Each person had a different amount of money and paid a different price for their item. Using the clues, the pupils have to write the given information on the recording sheet and then calculate the missing information.

Instructions

1 Pupil A has clue sheet A, Pupil B has clue sheet B, Pupil C has clue sheet C and Pupil D has clue sheet D. If there are only two pupils, they should have two clue sheets each. Each pupil has a copy of the recording sheet.

2 Pupils take turns to read a clue from their sheet and decide together where the information should be written on the recording sheet.

3 When they have correctly written all the information from the clues on the recording sheet, the pupils work together to calculate the missing information in each row.

Note: The statements in the clues and the question headings on the recording sheet are typical of the kind of language forms used in one-step word problems involving a change (reduction) from an initial number or amount to a result number or amount, that is initial set – change set – result set. The question can be about either the initial set number, the change set number or the result set number.

Substitution tables: Addition and subtraction

See download for full-size printable tables.

Content: Writing word problems and making them solvable.

Key language

See substitution tables.

Context

Pupils should understand that people and objects should remain consistent in one-step word problems and the numbers should 'work'. For example, the initial set quantity (Alex had 54 apples) should be greater than the change set quantity if there is a decrease in the initial set (Alex gave 14 apples to his friend Josef). Also, pupils need to keep the objects (apples, etc.) consistent in each word problem.

Asking questions

Use the table to create five word problems. Work out the answers. Give them to your friend to solve. Check your friend's answers to see if they are correct.

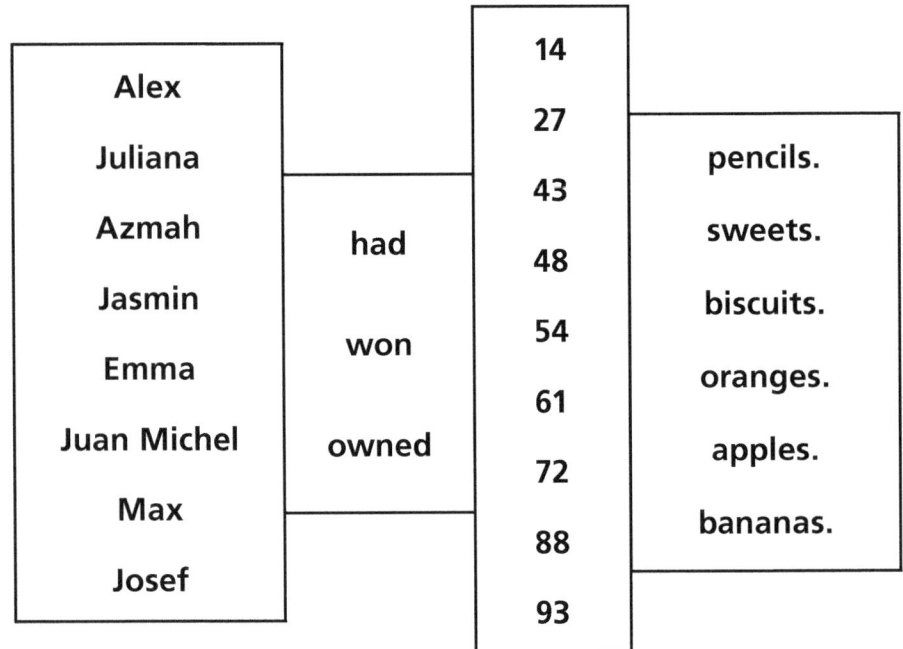

Alex Juliana Azmah Jasmin Emma Juan Michel Max Josef	had won owned	14 27 43 48 54 61 72 88 93	pencils. sweets. biscuits. oranges. apples. bananas.

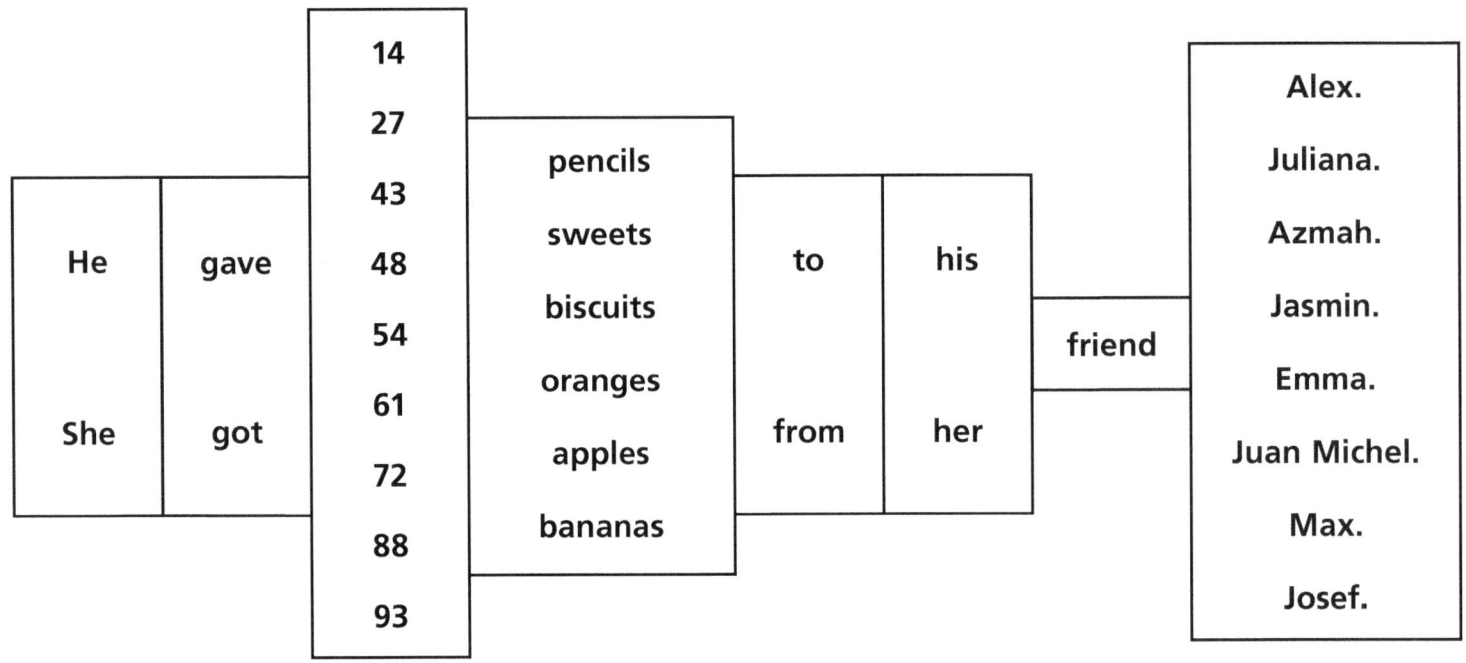

He She	gave got	14 27 43 48 54 61 72 88 93	pencils sweets biscuits oranges apples bananas	to from	his her	friend	Alex. Juliana. Azmah. Jasmin. Emma. Juan Michel. Max. Josef.

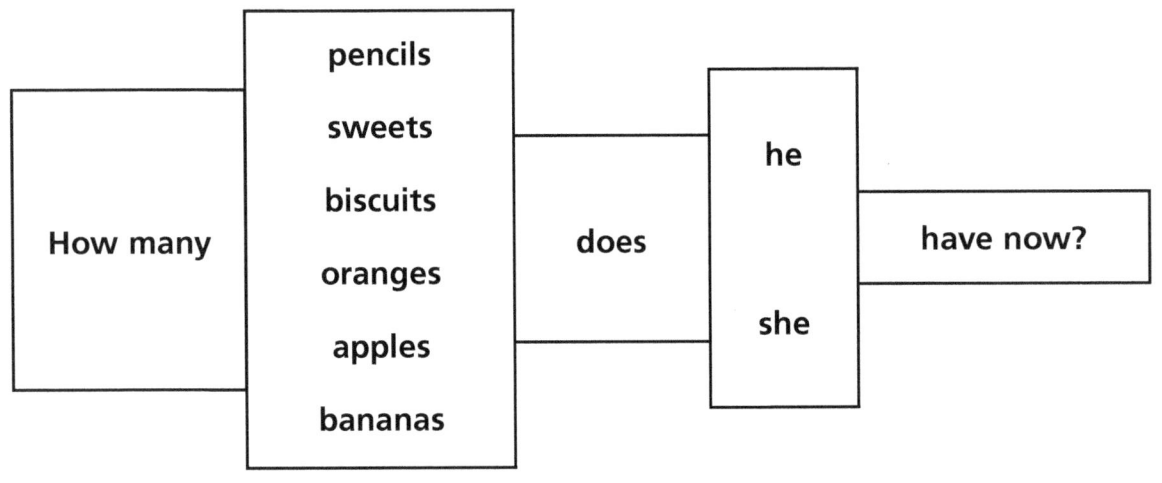

How many	pencils sweets biscuits oranges apples bananas	does	he she	have now?

Asking more questions

Write five more word problems. Work out the answers. Give them to your friend to solve. Check your friend's answers to see if they are correct.

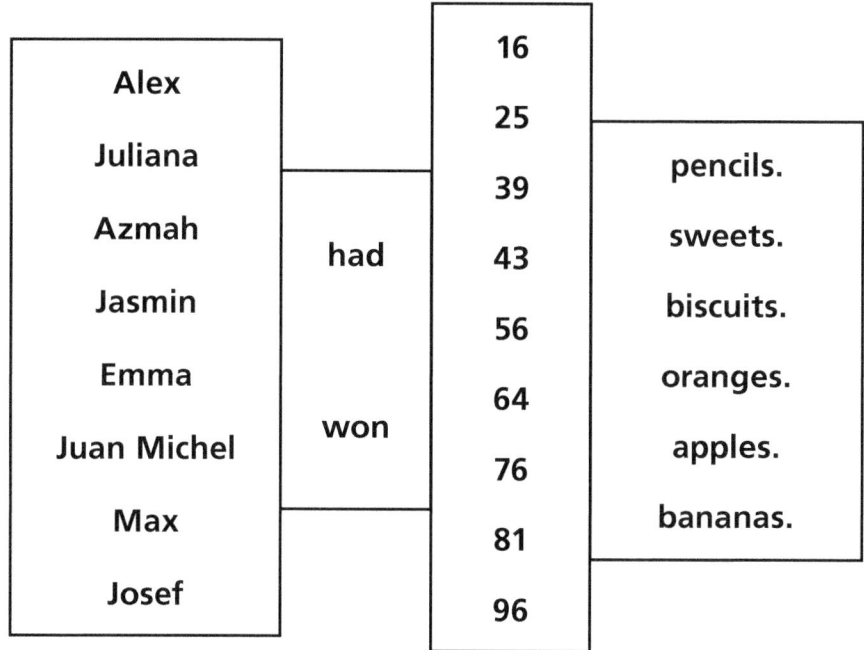

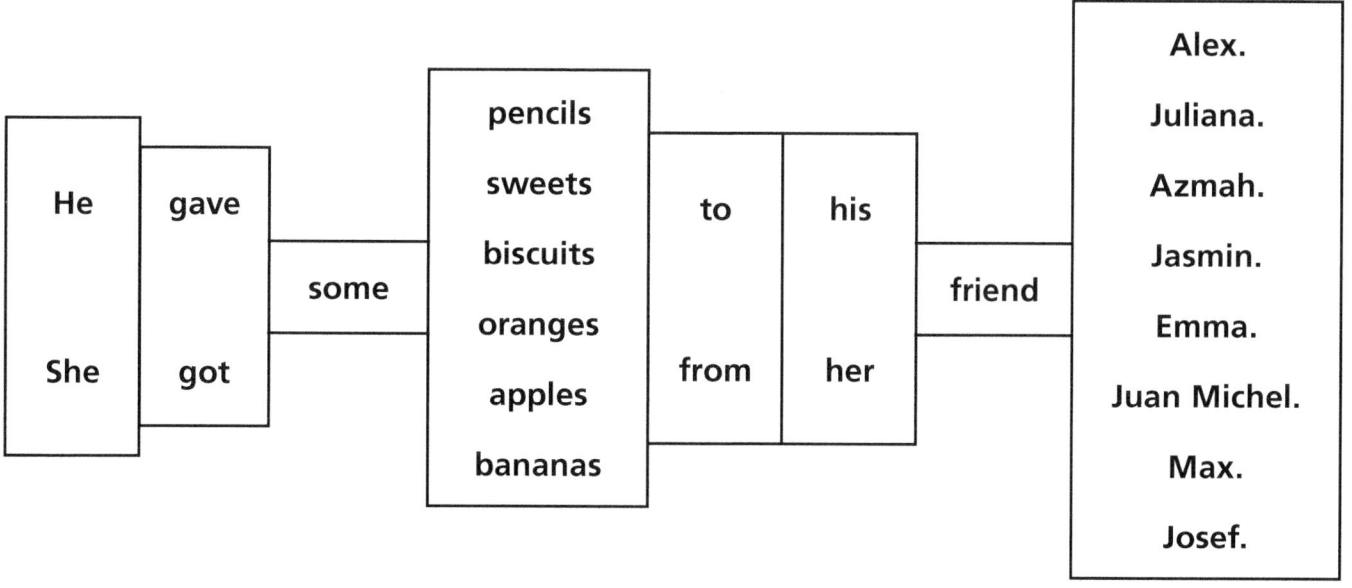

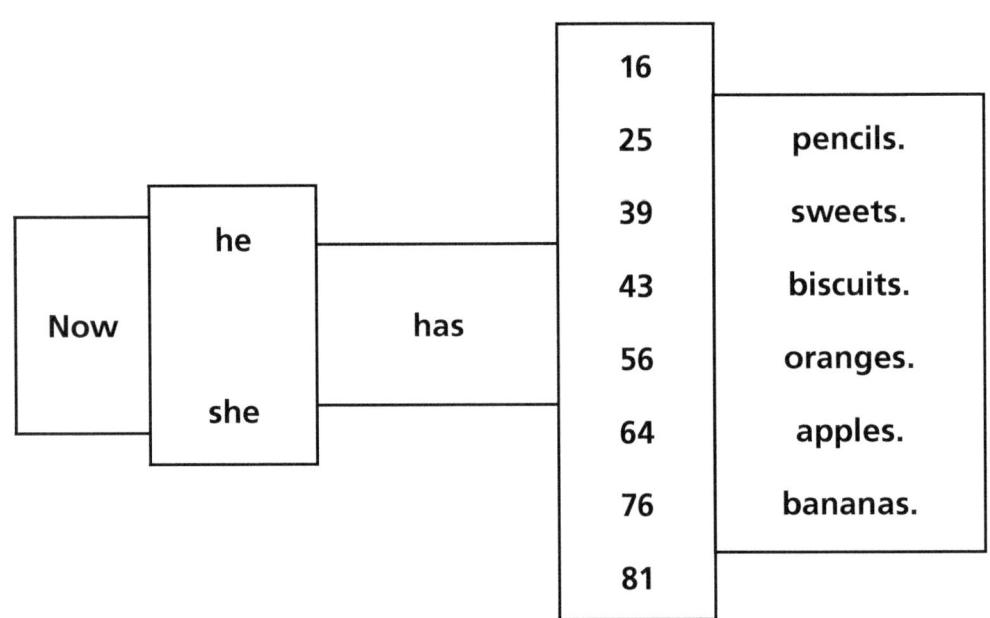

| Now | he / she | has | 16 / 25 / 39 / 43 / 56 / 64 / 76 / 81 | pencils. / sweets. / biscuits. / oranges. / apples. / bananas. |

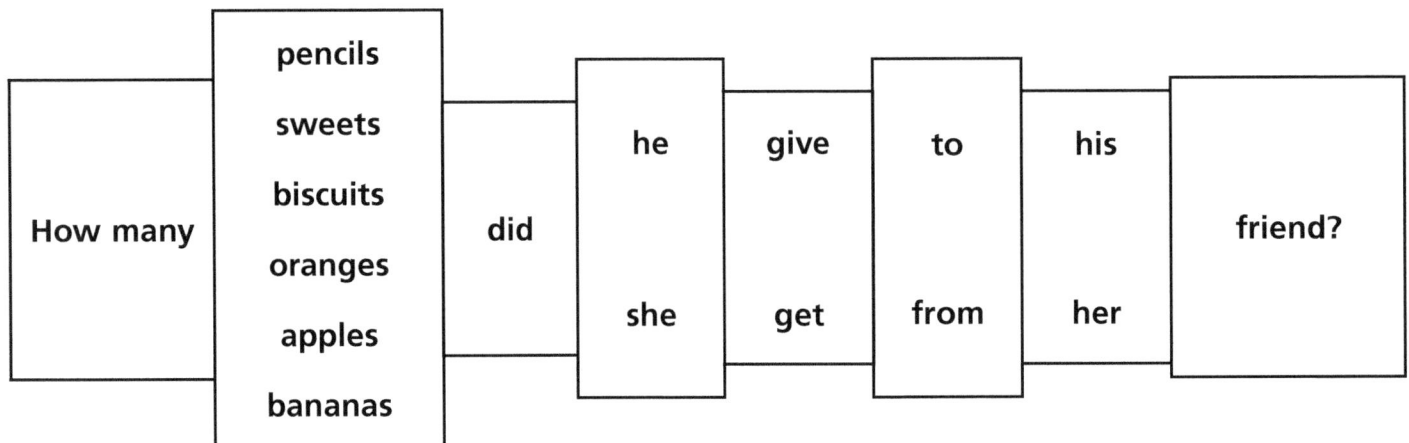

| How many | pencils / sweets / biscuits / oranges / apples / bananas | did | he / she | give / get | to / from | his / her | friend? |

Number – addition and subtraction

Questions with support

1 Kinga had £154. She spent £87 at the supermarket. How much money did she have left?

£ [ ]

> ! What numbers do you know from the question that you can write in the boxes below? What number do you not know? Can you calculate the unknown number? Do you have to add or subtract to calculate the unknown number?
>
> | In the first place, she had | → | Then she spent | → | After that she had |

2 Patryk had 366 football cards. His friends gave him 243 more cards. How many cards does Patryk have now?

[ ]

> ! What numbers do you know from the question that you can write in the boxes below? What number do you not know? Can you calculate the unknown number? Do you have to add or subtract to calculate the missing number?
>
> | In the first place, he had | → | Then | → | Now |

3 On Monday morning, a shop had 274 packets of sweets. On Tuesday morning, the shop had 139 packets left. How many packets did the shop sell on Monday?

[ ]

> ! What numbers do you know from the question that you can write in the boxes below? What number do you not know? Can you calculate the unknown number? Do you have to add or subtract to calculate the missing number?
>
> | On Monday, the shop had | → | The shop sold | → | On Tuesday morning, |

4 Sabrina read 187 pages of a book. Kamil read 98 pages. How many more pages did Sabrina read than Kamil?

```
┌────────────────┐
│                │
│                │
└────────────────┘
```

> What numbers do you know from the question that you can write in the boxes below? What number do you not know? Can you calculate the missing number? Do you have to add or subtract to calculate the missing number?

```
┌────────────────┐
│ Who read       │
│ more pages?    │
│                │
│ How many?      │
└───────┬────────┘                ┌──────────────────┐
        │         ──────────────▶ │ How many         │
        │                         │ more pages?      │
┌───────▼────────┐                │                  │
│ Who read fewer │                └──────────────────┘
│ pages?         │
│                │
│ How many?      │
└────────────────┘
```

5 Azar and Yusuf had 275 books altogether. Azar had 145 books. How many books did Yusuf have?

```
┌────────────────┐
│                │
│                │
└────────────────┘
```

> What numbers do you know from the question that you can write in the boxes below? What number do you not know? Can you calculate the missing number? Do you have to add or subtract to calculate the missing number?

```
┌──────────────┐              ┌──────────────┐
│ Azar had     │              │ Yusuf had    │
│              │              │              │
└──────┬───────┘              └──────┬───────┘
       │                             │
       └──────────┐      ┌───────────┘
              ┌───▼──────▼───────┐
              │ Altogether Azar and │
              │ Yusuf had           │
              └─────────────────────┘
```

6 Kaspar had 89 stickers. His friend gave him some more stickers. Now Kaspar has 147 stickers. How many stickers did his friend give Kaspar?

```
┌────────────────┐
│                │
│                │
└────────────────┘
```

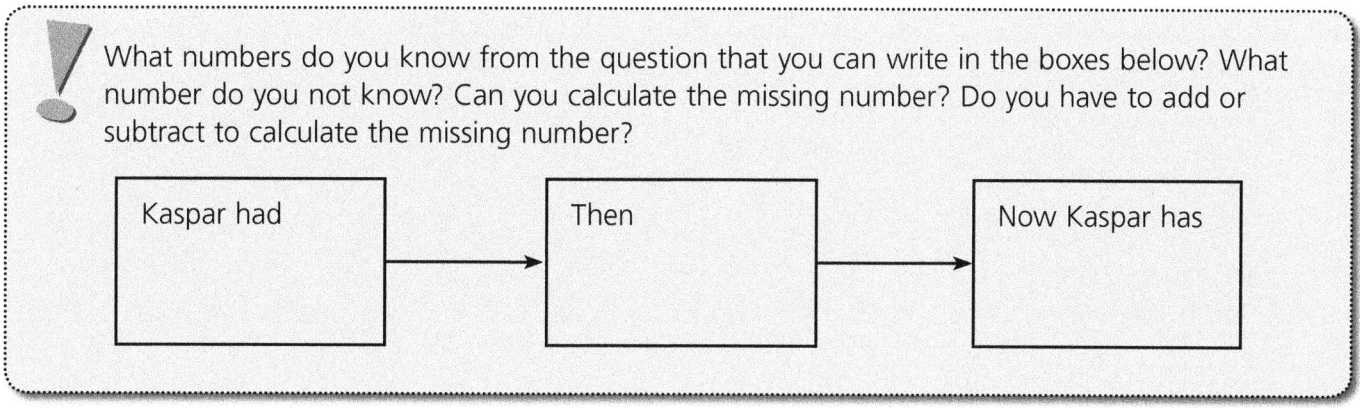

7 Cornel had £87. He wanted to buy a bicycle. The bicycle cost £156. How much more money did he need to buy the bicycle?

£

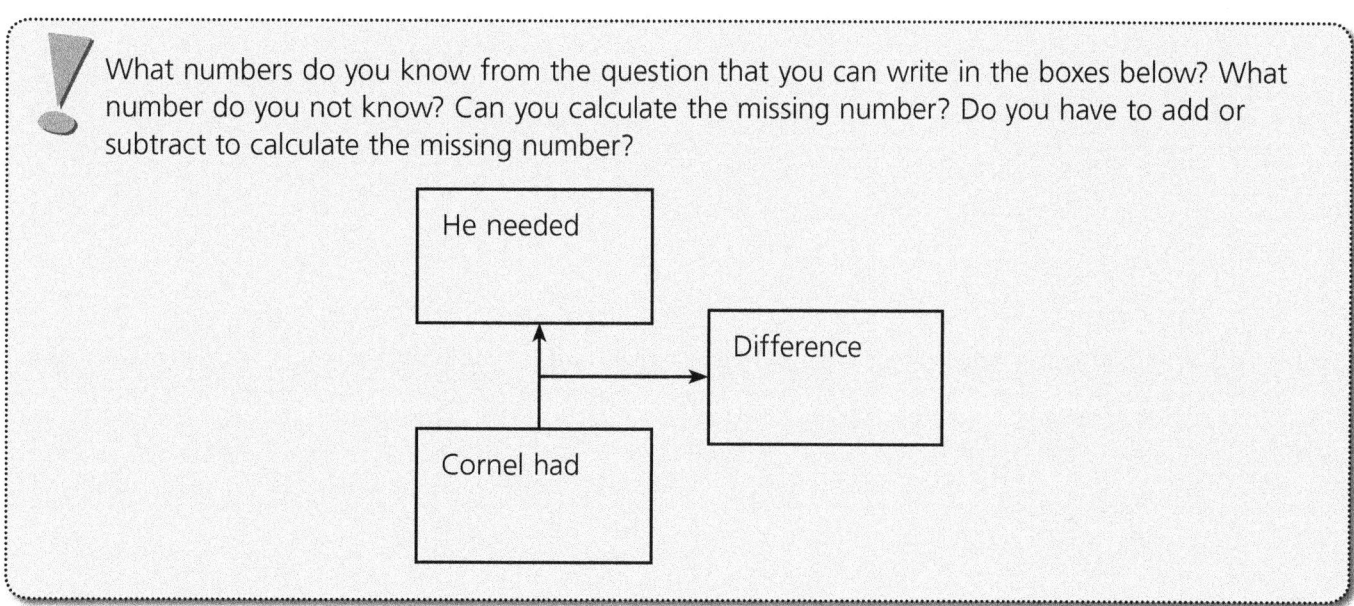

8 Seema had some photos on her phone. Then she went on holiday and took 149 more photos. Now she has 308 photos on her phone. How many photos did she have on her phone before she went on holiday?

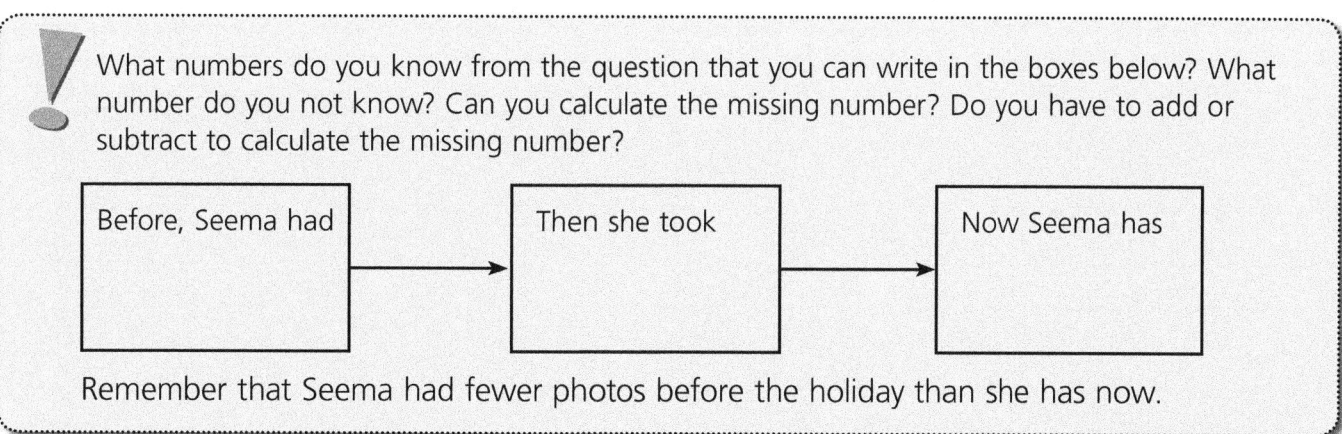

9 Andrea scored 234 points in a game. She scored 59 more points than her brother. How many points did her brother score?

┌─────────────────────┐
│   │
│   │
└─────────────────────┘

> What numbers do you know from the question that you can write in the boxes below? What number do you not know? Can you calculate the missing number? Do you have to add or subtract to calculate the missing number?
>
> ┌─────────────────┐
> │ Andrea scored   │
> │   │
> └─────────────────┘
>
> ┌──────────────────┐
> │ How many more   │
> │ points?   │
> │   │
> └──────────────────┘
>
> ┌─────────────────┐
> │ Her brother   │
> │ scored   │
> │   │
> └─────────────────┘
>
> Remember that her brother scored fewer points than Andrea.

10 Yusuf had these four digit cards.

┌───────┐ ┌───────┐ ┌───────┐ ┌───────┐
│   7   │ │   3   │ │   2   │ │   4   │
└───────┘ └───────┘ └───────┘ └───────┘

a. He made a number with two of the digit cards. He added the number to 51. The total was 83. What number did he make?

		+	5	1	=	8	3

> What number added to 1 one gives you a total of 3 ones?
>
> What number added to 5 tens gives you a total of 8 tens?

b. He then made a three-digit number using the digit cards. This time he subtracted 312 from the number. The answer was 412. What was the number?

			−	3	1	2	=	4	1	2

> What number when you subtract 2 ones from it gives you the answer 2 ones?
>
> What number when you subtract 1 ten from it gives you the answer 1 ten?
>
> What number when you subtract 3 hundreds from it gives you the answer 4 hundreds?

11 What are the missing digits in this column addition?

	hundreds	tens	ones
	3	4	5
+		6	
	8		9

What number when added to 5 gives you a total of 9?

What is the total if you add 4 and 6? Do you have to carry a digit to the hundreds column?

What number gives you a total of 8 when you add 3 and the carried digit?

In another example:

	hundreds	tens	ones
	2	9	3
+		2	
	7		8

When you add **5** to 3 it gives you a **total** of 8.

If you add 9 and 2 the **total** is **11**. You have to **carry** a digit to the hundreds column.

When you add **4** to 2 and the **carried digit** it gives you a total of 7.

	hundreds	tens	ones
	2₁	9	3
+	**4**	2	**5**
	7	**1**	8

12 **a.** What is the missing number in this bar model?

457	
137	

You can use the **inverse** to find **the missing number**, as in this example.

616	
342	

If you **subtract** 342 from the **total**, you find **the missing number**.

616 **minus** 342 equals 274. Therefore, **the missing number** is 274.

You can check this by **adding** 342 to 274. 342 **plus** 274 equals 616.

b. Write the missing number in the box.

	+	256	=	495

13 The table shows the weight of four types of shark.

Type of shark	Weight in kilograms
Tiger shark	635
Bull shark	130
Lemon shark	183
Mako shark	790

a. What is the difference in weight between the Tiger shark and the Lemon shark?

kg

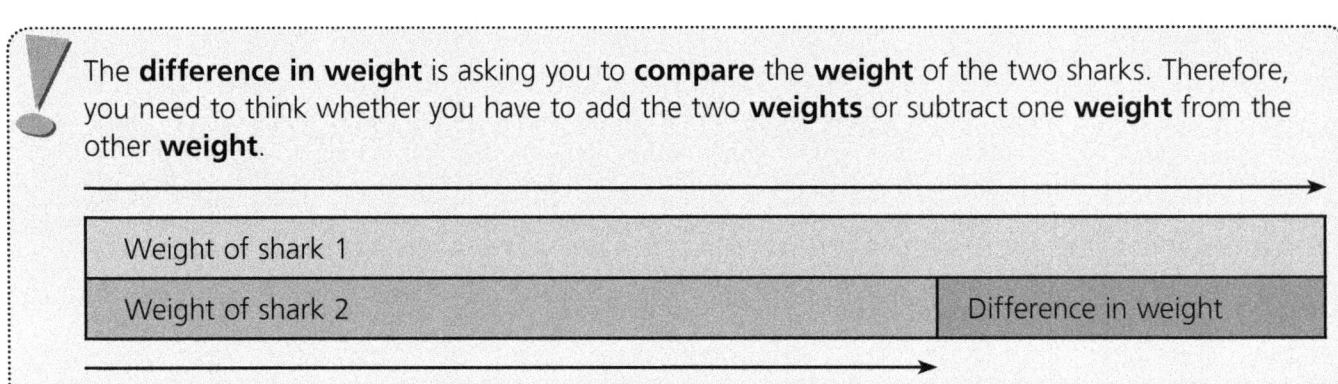

The **difference in weight** is asking you to **compare** the **weight** of the two sharks. Therefore, you need to think whether you have to add the two **weights** or subtract one **weight** from the other **weight**.

Weight of shark 1	
Weight of shark 2	Difference in weight

b. What is the difference in weight between the Mako shark and the combined weights of the Bull shark and the Lemon shark?

kg

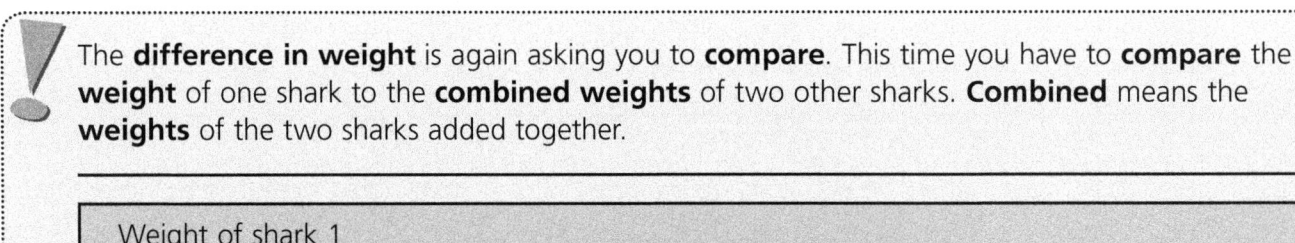

The **difference in weight** is again asking you to **compare**. This time you have to **compare** the **weight** of one shark to the **combined weights** of two other sharks. **Combined** means the **weights** of the two sharks added together.

Weight of shark 1		
Weight of shark 2	Weight of shark 3	Difference in weight
Combined weight of sharks 2 and 3		

Answers

1. £67

2. 609

3. 135

4. 89

5. 130

6. 58

7. £69

8. 159

9. 175

10. **a.** 32   **b.** 724

11.

	hundreds	tens	ones
	3$_1$	4	5
+	**4**	6	**4**
	8	**0**	9

12. **a.** 320   **b.** 239

13. **a.** 452kg   **b.** 477kg

Number – multiplication and division

Collaborative activities

Information gap / barrier game: 3, 4 and 8 multiplication tables

An activity for two pupils. *See download for accompanying printable resources.*

Content: Matching multiplication cards with the same answer

Key language

Four multiplied by four has the **same answer as eight multiplied by two**.

Instructions

1. Pupil A takes the grey board and the set of cut-up grey cards. Pupil B takes the white board and the set of cut-up white cards.
2. Each pupil places cards over the squares that match on their board. Each pupil will be left with four cards and now needs to find out where they go on their board.
3. Pupil A says: "I need to know what goes in the first square in the right-hand column."
4. Pupil B says: "8 × 2. Which card contains a calculation that gives you the same answer?"
5. Pupil A works out that 4 × 4 and 8 × 2 give the same answer and says: "I think that 4 × 4 gives the same answer." If pupil B agrees, the card can be placed in the empty square.
6. The roles are then reversed.
7. Each pupil plays in turn until both boards are completed.

Four in a line: Multiplication

An activity for two pupils, or four pupils in two teams of two. *See download for accompanying printable resources.*

Content: Multiplication

Key language

Three **multiplied by** five equals fifteen.

Instructions

1. Cut out the cards.
2. One pupil or team has the yellow cards and the other pupil or team has the green cards.
3. Place the cards face down in two piles.
4. The yellow team turns over a card and places it on the box which is the answer to the calculation on the card, saying aloud why the card goes on that box (e.g. "Three multiplied by five equals fifteen.").
5. The green team turns over a card and places it on the box which is the answer to the calculation on the card, saying aloud why the card goes on that box (e.g. "Four multiplied by two equals eight.").
6. The aim of the game is to get four cards of your colour in a line – across, down or diagonally.
7. The winner is the first pupil or team to get four of their colour in a line.

Domino chains: Division

An activity for the whole class. *See download for accompanying printable resources.*

Content: Division

Key language

What is forty-four **divided by** four? What is thirty **divided by** three?

Instructions

1. Cut out the dominoes. Each domino has a whole number on the left (e.g. 12) and a division calculation on the right (e.g. 44 ÷ 4).

2. Give a domino to each pupil. There are 33 dominoes in total so you may need to give some pupils more than one. Tell the pupils that they will play all of the yellow dominoes first, then all of the green dominoes, then all of the blue dominoes.

3. Choose a pupil with a yellow domino to read out their calculation (e.g. "What is forty-four divided by four?").

4. The pupil with the yellow domino containing the answer on the left-hand side says: "I have the answer. It is 11. The calculation on my domino is forty divided by four."

5. The pupil with the yellow domino containing the answer to this calculation says: "I have the answer. It is 10. The calculation on my domino is twenty-seven divided by three."

6. Repeat until all the dominoes have been used.

Substitution tables: Division

See download for full-size printable tables.

Content: Making statements about division

Key language

See substitution tables.

Making statements

Use the table to create 10 correct statements. You will need to use some numbers from each column more than once. Write each division in words, and then, underneath, write the division in numerals. For example:

Twenty-four divided by 8 equals 3

24 ÷ 8 = 3

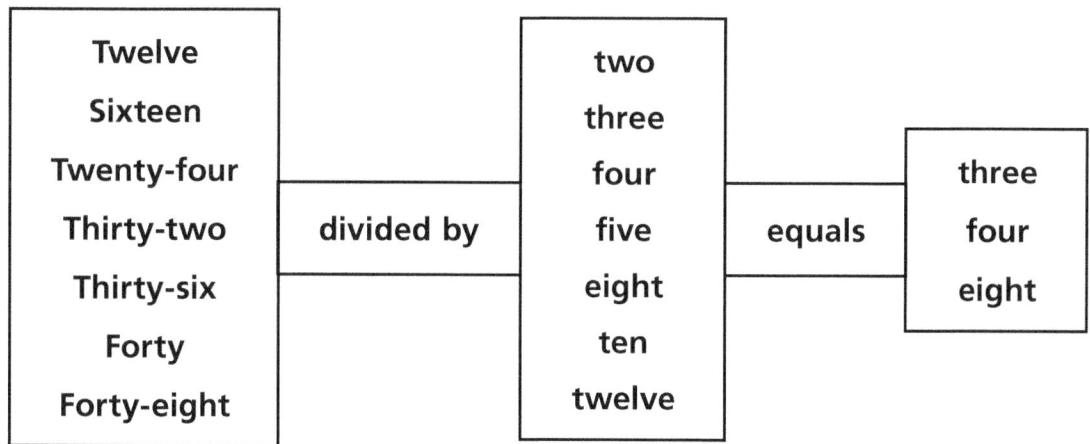

Twelve		**two**		
Sixteen		**three**		**three**
Twenty-four		**four**		
Thirty-two	**divided by**	**five**	**equals**	**four**
Thirty-six		**eight**		
Forty		**ten**		**eight**
Forty-eight		**twelve**		

Number – multiplication and division

Questions with support

1 Write the missing numbers to complete the multiplication grid.

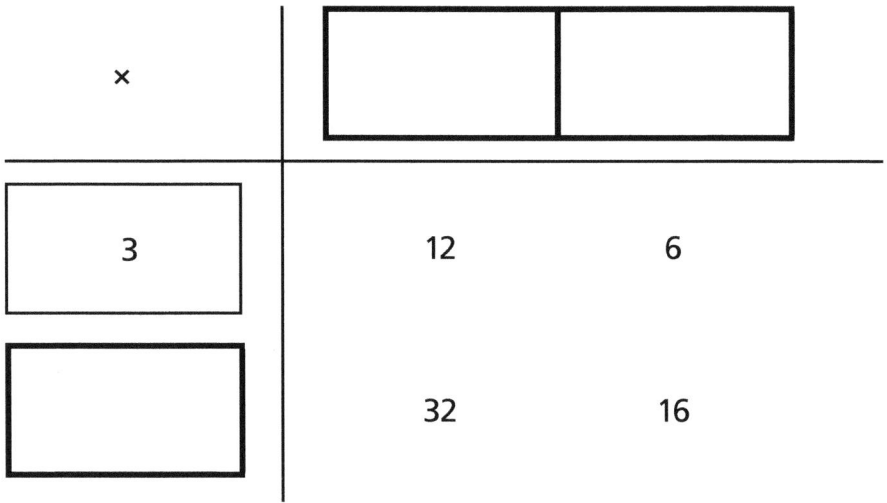

Write the missing numbers to complete the multiplication grid.

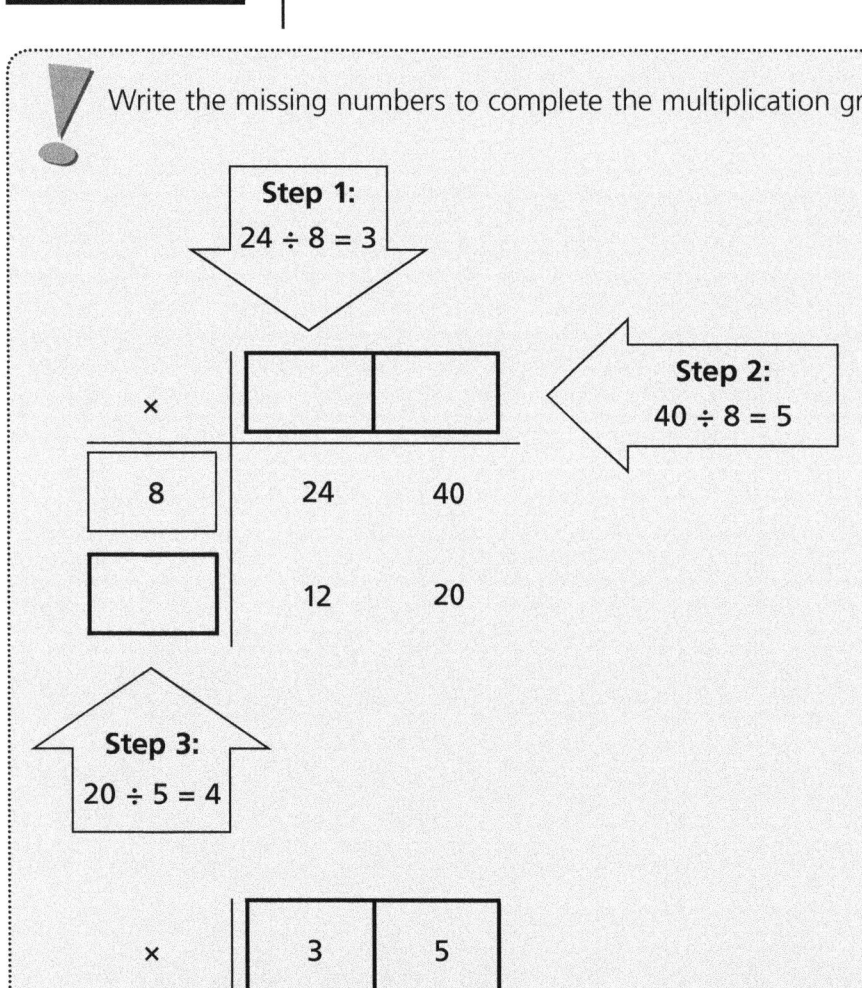

2 Anna uses these digit cards.

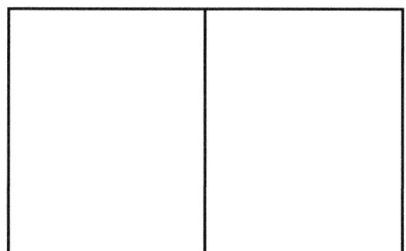

She makes a two-digit number and a one-digit number.

She multiplies them together. Her answer is a multiple of 8.

What could Anna's multiplication be?

 Adam uses these digit cards.

6 | 1 | 2

He makes a two-digit number and a one-digit number.

He multiplies them together. His answer is a multiple of 8. What **could** Adam's multiplication be?

61 × 2 = 122   21 × 6 = 126   12 × 6 = 72 ✓

26 × 1 = 26   16 × 2 = 32 ✓   62 × 1 = 62

The answer could be 12 × 6 = 72 or 16 × 2 = 32.

Remember: **could** means there might be more than one correct answer.

3 A large bag of balloons contains 16 balloons. A small bag contains half as many balloons as a large bag. Maria has a small bag and a large bag. How many balloons does she have altogether?

> Large bag
>
> 16 balloons

> Small bag
> of balloons

[ ]

> Remember, **a large bag contains** means a large bag **has**.
>
> A large bag of marbles contains 12 marbles. A small bag contains half as many marbles as a large bag. Adam has a small bag and a large bag. How many marbles does he have altogether?
>
> > Large bag
> >
> > 12 marbles
>
> > Small bag
> > of marbles
>
> The small bag has **half as many marbles** as the large bag.
>
> $12 \div 2 = 6$
>
> **Add** the 6 marbles from the small bag to the 12 marbles from the large bag.
>
> $12 + 6 = 18$

4 Anita buys an ice-cream for £2. How much does it cost to buy 3 ice-creams?

£ [ ]

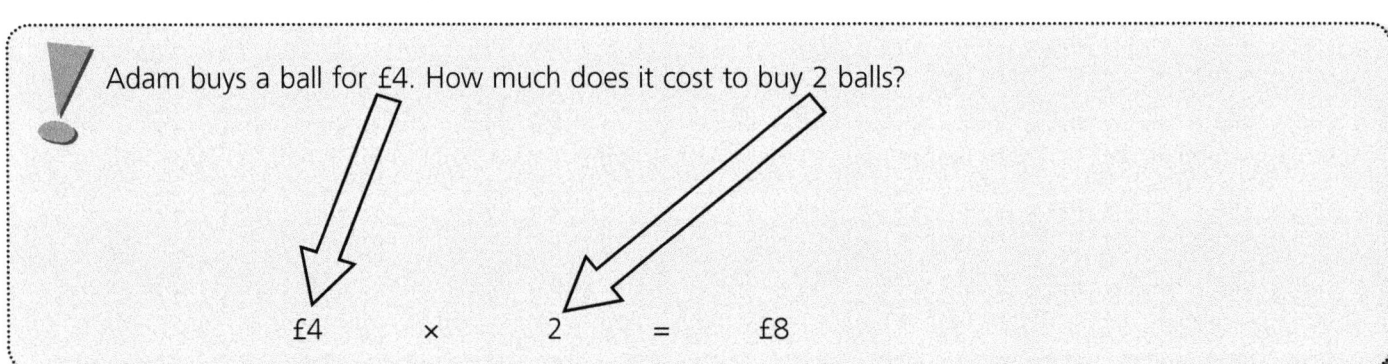

> Adam buys a ball for £4. How much does it cost to buy 2 balls?
>
> £4 × 2 = £8

 Fill in the missing numbers to complete the sequence.

	15		21	24		30

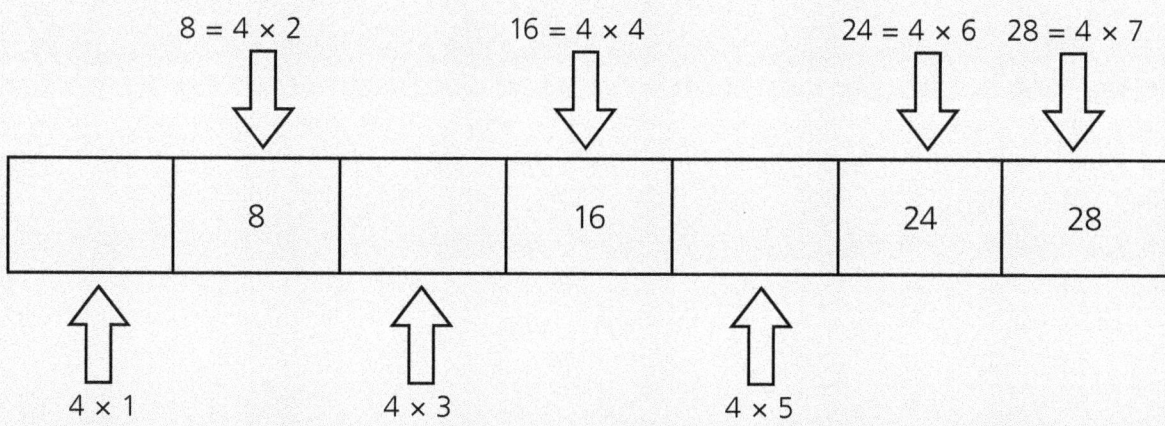

To answer this question, you need to work out which multiplication table the given numbers belong to.
For example:

Fill in the missing numbers to complete the sequence.

	8		16		24	28

8 = 4 × 2   16 = 4 × 4   24 = 4 × 6   28 = 4 × 7

4 × 1   4 × 3   4 × 5

Therefore, the answer is:

4	8	**12**	16	**20**	24	28

Answers

1.

×	**4**	**2**
3	12	6
8	32	16

2.

1	2	×	4

or

2	4	×	1

3. 24

4. £6

5.

12	15	**18**	21	24	**27**	30

Number – fractions

Collaborative activities

Four in a line: Fractions

An activity for two pupils, or four pupils in two teams of two. *See download for accompanying printable resources.*

Content: Reading fractions as numbers and words

Key language

Four-fifths is written as $\dfrac{4}{5}$

Seven-eighths is written as $\dfrac{7}{8}$

Instructions

Note: The board and sheet of cards should be enlarged to A3 size. There are two versions of the game – it can be played using the words board and number cards or using the numbers board and the word cards.

1. Cut the cards into individual cards.
2. One pupil or team has the yellow cards and the other pupil or team has the green cards.
3. Place the cards face down in two piles.
4. The yellow team takes the top card from their pile and places it on a correct box on the board. When they put the card down, they must say a correct sentence (e.g. "$\dfrac{4}{5}$ is written as four-fifths").
5. The green team then places their card and says a sentence.
6. The aim of the game is to get four cards of your colour in a line – across, down or diagonally.
7. The winner is the first pupil or team to get four of their colour in a line.

Card matching: Reading fractions

An activity for two, three or four pupils. *See download for complete printable activity.*

Content: Reading fractions

Key language

$\dfrac{4}{5}$ is read as **four-fifths.**

$\dfrac{1}{3}$ is read as **one-third.**

$\dfrac{4}{7}$ is read as **four-sevenths.**

Instructions

1. Cut the cards into individual cards.
2. Share out the cards among the pupils.
3. The pupils place the cards containing fractions written in numbers on the board.
4. They then take turns to read aloud one of their fractions in words cards and match the card to the correct fraction in numbers card on the board.
5. When they have matched all the cards, each pupil uses their own copy of the board to write the fractions and words.

Information gap / barrier game: Shaded circles and fractions

An activity for two pupils. *See download for complete printable activity.*

Content: Finding fractions of a circle

Key language

Tell me **what fraction of my circle to shade**.

The circle in the first column on the top row **has one quarter shaded**.

Context

Pupils share information to complete a grid of fractions or shaded circles.

Instructions

1. Pupil A has sheet A and Pupil B has sheet B.
2. Pupil A has information on their sheet which Pupil B does not have and Pupil B has information which Pupil A does not have.
3. Pupils ask each other questions to fill in their missing information. Pupil A begins by saying: "In the first column on the top row, I have a circle with four quarters. How many should I shade?" Pupil B responds: "Shade one quarter. What fraction of the circle in the second column of the top row is shaded?" Pupil A shades one quarter of the circle then responds to the question: "Five-eighths." Pupil B writes the fraction as numbers on their sheet. Pupil A says: "Now please tell me how many fifths of the next circle on the top row I should shade."
4. Each pupil plays in turn until both sheets are complete. Repeat the activity with Pupil A using sheet B and Pupil B using sheet A.

Substitution tables: Fractions

See download for full-size printable tables.

Content: Writing fractions as words and numbers

Key language

See substitution tables

Making sentences

In your pair, use the substitution table to create nine correct sentences. You can only use each word or fraction once. You cannot write a sentence unless you both agree that it is correct.

one	ninths		$\frac{5}{6}$
two	fifths		$\frac{9}{10}$
three	sixths		$\frac{4}{5}$
four	quarters		$\frac{6}{7}$
five	half	is written as	$\frac{1}{2}$
six	tenths		$\frac{8}{9}$
seven	eighths		$\frac{7}{8}$
eight	sevenths		$\frac{3}{4}$
nine	thirds		$\frac{2}{3}$

Making more sentences

Use the table to create nine correct sentences of your own. You can use words more than once. You do not have to use all the words. Ask your friend to check them to make sure they are true.

one			
two	ninth/s		
three	fifth/s		
four	sixth/s		
five	quarter/s	is written as	
six	tenth/s		
seven	eighth/s		
eight	seventh/s		
nine	third/s		

Number – fractions

Questions with support

1 Tick (✓) two shapes that have $\frac{1}{4}$ shaded.

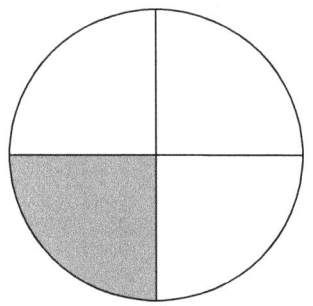 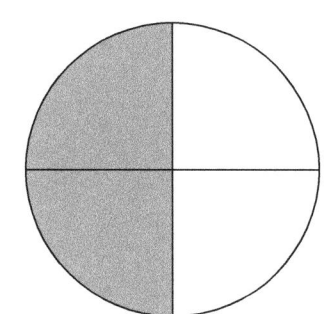 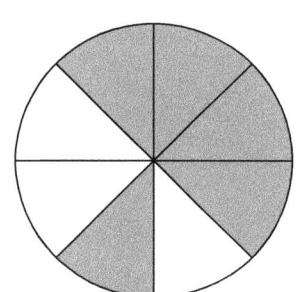 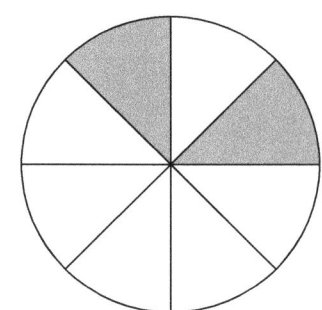

 Tick (✓) two shapes that have $\frac{1}{2}$ shaded.

✓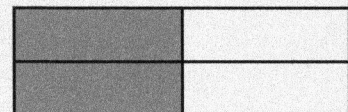

$$\frac{2}{4} = \frac{1}{2}$$

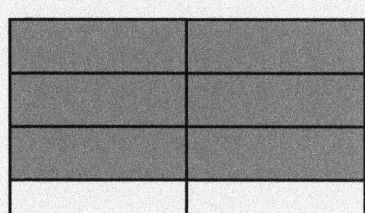

$$\frac{6}{8} = \frac{3}{4}$$

✓

$$\frac{3}{6} = \frac{1}{2}$$

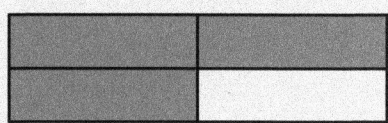

$$\frac{3}{4}$$

2 Put the fractions in the correct order from the smallest to the largest.

$\frac{1}{8}$ 　　　　 $\frac{1}{4}$ 　　　　 $\frac{1}{3}$ 　　　　 $\frac{1}{5}$

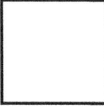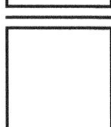 　  　 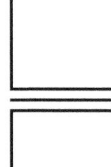 　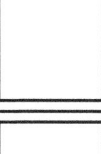

Smallest 　　　　　　　　　　　　　　　　　 **Largest**

$\dfrac{1}{4}$ $\qquad$ $\dfrac{1}{7}$ $\qquad$ $\dfrac{1}{5}$ $\qquad$ $\dfrac{1}{2}$

1	1	1	1
7	5	4	2

Smallest $\qquad\qquad\qquad\qquad\qquad\qquad\qquad\qquad\qquad\qquad$ **Largest**

3 Put the fractions in the correct order from the smallest to the largest.

$\dfrac{3}{7}$ $\qquad$ $\dfrac{6}{7}$ $\qquad$ $\dfrac{2}{7}$ $\qquad$ $\dfrac{4}{7}$

Smallest $\qquad\qquad\qquad\qquad\qquad\qquad\qquad\qquad\qquad\qquad$ **Largest**

$\dfrac{2}{6}$ $\qquad$ $\dfrac{5}{6}$ $\qquad$ $\dfrac{3}{6}$ $\qquad$ $\dfrac{4}{6}$

2	3	4	5
6	6	6	6

Smallest $\qquad\qquad\qquad\qquad\qquad\qquad\qquad\qquad\qquad\qquad$ **Largest**

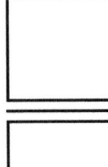 Adam has 5 apples and Chen has 4 apples and 1 banana. What fraction of their fruits are apples? Write your answer in the boxes.

$$\frac{}{}$$

 Anna has 5 red balloons and Ali has 3 blue balloons and 2 red balloons. What fraction of **their** balloons are red? Write your answer in the boxes.

Their tells you the question is about how many balloons Anna and Ali have altogether and what fraction of them are red.

Red balloons: 5 + 2 = 7

Balloons: 5 + 2 + 3 = 10

Fraction of balloons that are red =

$$\frac{7}{10}$$

5 Complete the final two boxes.

$$\frac{3}{8} + \frac{2}{8} = \frac{}{}$$

 **Example**
Complete the final two boxes.

$$\frac{4}{7} + \frac{2}{7} = \frac{}{}$$

Remember, the **denominator** is the number of equal parts the whole is split into and it is the number below the line. The **numerator** is the number above the line and is the number of equal parts in the fraction. If the number below the line is the same, you can add and subtract the numbers above the line.

$$\frac{4}{7} + \frac{2}{7} = \frac{6}{7}$$

6 There are 9 balls in a box.

Anna, Chen and Ali each take 2 balls out of the box. Fiona takes 1 ball out of the box.
What fraction of the balls are left in the box?

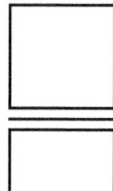

There are 8 balls in a box.

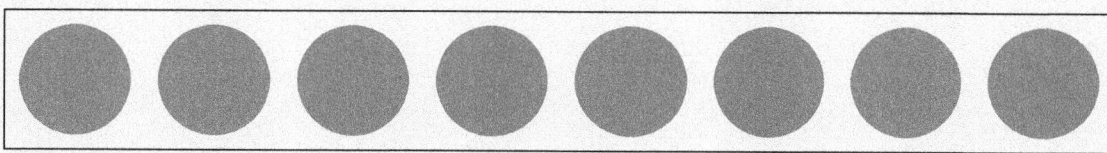

Anna, Chen and Ali each take 1 ball out of the box. Fiona takes 2 balls out of the box.
What fraction of the balls are left in the box?

Anna		Chen		Ali		Fiona		Total balls taken out of the box
$\frac{1}{8}$	+	$\frac{1}{8}$	+	$\frac{1}{8}$	+	$\frac{2}{8}$	=	$\frac{5}{8}$

All balls		Total balls taken out of the box		Balls left in the box
$\frac{8}{8}$	−	$\frac{5}{8}$	=	$\frac{3}{8}$

Answers

1.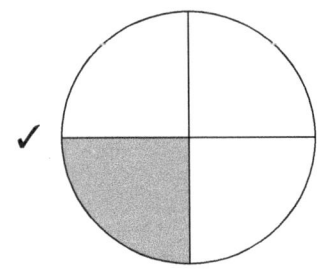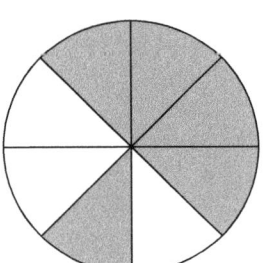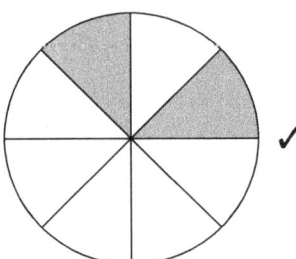

2. $\frac{1}{8}$ $\frac{1}{5}$ $\frac{1}{4}$ $\frac{1}{3}$

3. $\frac{2}{7}$ $\frac{3}{7}$ $\frac{4}{7}$ $\frac{6}{7}$

4. $\frac{9}{10}$

5. $\frac{5}{8}$

6. $\frac{2}{9}$

Explain why

Collaborative activities

Card sorting: Laura's fruit

An activity for two, three or four pupils. *See download for complete printable activity.*

Content: Expressing and explaining possibilities and logical necessities about the possible combinations of three kinds of fruit

Key language

She **could have** five oranges **because** she **could have** four apples and two bananas.

She **can't have** six apples **because** she **can't have** more apples than oranges.

She **can't have** 10 oranges **because** she **must have at least** two apples and one banana, which is more than 11 fruits.

She **must have more** apples **than** bananas.

Context

Laura has 11 pieces of fruit. She has three kinds of fruit: oranges, apples and bananas. She has more oranges than apples. She has more apples than bananas. Pupils work out what combinations of fruits Laura could have.

Instructions

This activity is intended as a guided talk in which the teacher or teaching assistant works with a small group of two, three or four pupils.

1 Give the pupils a copy of the sorting board and a set of number cards. It is useful if pupils are provided with plastic fruit or coloured counters to represent the different kinds of fruit to support them with the activity.

2 The pupils use the number cards to work out a combination of the three fruits which amount to the total of 11. They place the number cards on the sorting board to show how many of each fruit there are in the combination. The combination must have more oranges than apples and more apples than bananas.

3 Pupils record the combination on their own copies of the board.

4 They then find another combination of the three fruits which amount to the total of 11. There are five combinations altogether. Each card should only be used once.

5 When pupils have finished, encourage them to verbalise the information from the sorting board by asking prompt questions. For example, the pupils may be prompted to say: "She could have eight oranges, two apples and one banana" or "She can't have four bananas.".

6 Pupils then use the information from the sorting board to complete the guided sentences in response to the questions.

Clue sheets: Calculating clothes prices

An activity for two pupils. *See download for complete printable activity.*

Content: Using subtraction to calculate the price of clothing items

Key language

Aziz paid £43 for a jacket **and** a shirt. The **price** of a jacket is £25. £43 **minus** £25 **equals** £18. **Therefore**, the price of a shirt **must be** £18.

Nine people bought different combinations of items of clothing which all had a different total price. The clues state how much each person paid for different combinations of items. Pupils calculate the prices of individual items and write explanations.

1. Pupil A has clue sheet A and Pupil B has clue sheet B. Each pupil has a copy of the recording sheet.

2. Pupils take turns to read a clue from their sheet and decide together where to write the information on the recording sheet so that how much the person paid is written in the correct row and column.

3. When they have recorded all the information from the clues, the pupils work together to calculate the missing information. Pupils may need reminding that they already know the price of a jacket.

4. The pupils then use the guided sentences to explain how the price of an item is calculated and why it is correct.

Note: The calculations are not difficult. The aim is to provide a sentence template which provides a way of explaining how you can calculate a single value from a total of two values. This can be used with larger numbers and combinations of distances, weights, times and other contexts.

Information gap / barrier game: Bar charts

An activity for four pupils. *See download for complete printable activity.*

Content: Comparing data from bar charts and explaining conclusions

Key language

How many children in Class X like football?

More children in Class X like football **than** like basketball.

It is true because nine children like football and only seven children like basketball. **Therefore, more** children in Class X like football **than** like basketball.

Context

The children in Class X and Class Y were asked what their favourite sport was. The bar charts for each class show how many children chose a particular sport as their favourite. Pupils ask and answer questions about the bar charts and respond orally and in writing to say whether statements are true or not.

Instructions

1. Pupil A has bar chart AX and Pupil B has bar chart BX. Both of these bar charts have data about Class X.

2. Pupil C has bar chart CY and Pupil D has bar chart DY. Both of these bar charts have data about Class Y.

3. Pupil A has information on their bar chart which Pupil B does not have and Pupil B has information which Pupil A does not have.

4. Pupil C has information on their bar chart which Pupil D does not have and Pupil D has information which Pupil C does not have.

5. Pupils ask each other questions to fill in their missing information (e.g. "How many children in Class X like football?"). The pupils shade the missing bars on their bar charts.

6. When they have exchanged all the information and shaded the missing bars on their bar charts, all the pupils work together to complete the guided sentences. The guided sentence sheet has three statements about Class X, three statements about Class Y and three statements that compare data about both classes. The pupils have to decide if the sentences are true or false. All four pupils have to respond to all the statements. Therefore, the pupils have to share the data from their bar charts through speaking and listening to each other.

Explain why

Questions with support

1 Diana is thinking of a number.

She says it is less than 12 and it is a multiple of 4.

a. Alex says that the number could be 6. Is Alex right?

Alex	is / isn't	right.		The number	could / can't	be	_____	because

_____	is / isn't	a multiple of / less than	4 / 12	and / but	it	is / isn't	a multiple of / less than	4. / 12.

b. Clare says that the number could be 8. Is Clare right?

Clare	is / isn't	right.		The number	could / can't	be	_____	because

_____	is / isn't	a multiple of / less than	4 / 12	and / but	it	is / isn't	a multiple of / less than	4. / 12.

2 The chart shows the number of excellent work points Class 3 got in week 1 and week 2. Each star equals 4 points.

☆ = 4 points

	Week 1	Week 2
Class 3	☆☆☆	☆☆☆☆☆

a. Sarah says that Class 3 got 2 more points in week 2 than in week 1. Is she right?

Sarah is / isn't right. Class 3 had _____ stars in week 1. _____ stars equals _____ points. In week 2, Class 3 had _____ stars. _____ stars equals _____ points. _____ minus _____ equals _____.

Therefore, Class 3 had _____ more points in week 2 than in week 1.

b. Class 4 had 16 more points than Class 3 in week 1. How many stars did Class 4 have on their chart in week 1?

Class 3 had _____ points in week 1. Class 4 had _____ more points than Class 3 in week 1. _____ plus _____ equals _____ points. One star equals _____ points. _____ divided by 4 equals _____.

Therefore, Class 4 had _____ stars on their chart in week 1.

3 These are the prices of three items for sale in a shop.

	Price
DVD	£16
Computer game	£12
Book	£5

a. Jason wants to buy a DVD, a computer game and a book. He has £30. Does he have enough money to buy all three things?

> Jason has / doesn't have enough money to buy a DVD, a computer game and a book because £_____
>
> plus £_____ plus £_____ equals £_____. Jason has £_____ and so he has / doesn't have enough money
>
> to buy all three things.

b. If Jason buys only the computer game, how much money will he have left?

> Jason has £_____. The computer game costs £_____. £_____ minus £_____ equals £_____.
>
> Therefore, he will have £_____ left.

4 Zahir has these three digit cards.

5	6	7

a. He says that the largest number he can make with these digit cards is 657. Is Zahir right?

> Zahir is / isn't right because with the digits 5, 6 and 7, he can make the number _____.
>
> _____ is more than _____ and so _____ is / isn't the largest number.

b. Zahir then says that 567 is less than 576. Is he right?

> Zahir is / isn't right because _____ has _____ hundreds and _____ tens. _____ has _____
>
> hundreds and _____ tens and so _____ is more than / less than _____.

5 What is wrong with this sequence?

50	100	150	250	300
first number	second number	third number	fourth number	fifth number

> The second number is _____ more than the first number. The third number is _____ more than
>
> the second number, but the fourth number is _____ more than the third number. The fourth
>
> number should be _____ because _____ is 50 / 100 more than _____. The fifth number should
>
> be _____ because _____ is 50 / 100 more than _____.

46

Answers

1. **a.** Alex **isn't** right. The number **can't** be **6** because **6 is less than 12 but** it **isn't a multiple of 4**

   or

   Alex **isn't** right. The number **can't** be **6** because **6 isn't a multiple of 4 but** it **is less than 12**.

   b. Clare **is** right. The number **could** be **8** because **8 is less than 12 and** it **is a multiple of 4.**

   or

   Claire **is** right. The number **could** be **8** because **8 is a multiple of 4 and** it **is less than 12**.

2. **a.** Sarah **isn't** right. Class 3 had **3** stars in week 1. **3** stars equals **12** points. In week 2, Class 3 had **5** stars. **5** stars equals **20** points. **20** minus **12** equals **8**. Therefore, Class 3 had **8** more points in week 2 than in week 1.

   b. Class 3 had **12** points in week 1. Class 4 had **16** more points than Class 3 in week 1. **16** plus **12** equals **28** points. One star equals **4** points. **28** divided by 4 equals **7**. Therefore, Class 4 had **7** stars on their chart in week 1.

3. **a.** Jason **doesn't have** enough money to buy a DVD, a computer game and a book because £**16** plus £**12** plus £**5** equals £**33**. Jason has £**30** and so he **doesn't have** enough money to buy all three things.

   b. Jason has £**30**. The computer game costs £**12**. £**30** minus £**12** equals £**18**. Therefore, he will have £**18** left.

4. **a.** Zahir **isn't** right because with the digits 5, 6 and 7, he can make the number **765**. **765** is more than **657** and so **657 isn't** the largest number.

   b. Zahir **is** right because **567** has **5** hundreds and **6** tens. **576** has **5** hundreds and **7** tens and so **567** is **less than 576**.

   or

   Zahir **is** right because **576** has **5** hundreds and **7** tens. **567** has **5** hundreds and **6** tens and so **576** is **more than 567**.

5. The second number is **50** more than the first number. The third number is **50** more than the second number, but the fourth number is **100** more than the third number. The fourth number should be **200** because **200** is **50** more than **150**. The fifth number should be **250** because **250** is **50** more than **200**.

Acknowledgments

Published by Keen Kite Books
An imprint of HarperCollins*Publishers* Ltd
The News Building
1 London Bridge Street
London
SE1 9GF

Text and design © 2018 Keen Kite Books, an imprint of HarperCollins*Publishers* Ltd
10 9 8 7 6 5 4 3 2 1
ISBN 978-0-00-822673-2

Authors: Graham Smith and Steve Cooke, The EAL Academy
Series Concept and Commissioning: Michelle I'Anson and Shelley Teasdale
Project Manager/Editor: Fiona Watson
Cover Design: Anthony Godber
Text Design and Layout: QBS Learning
Production: Natalia Rebow

A CIP record of this book is available from the British Library.